Life Science

___ Grade 2 ___

Written by Tracy Bellaire

The experiments in this book fall under ten topics that relate to two aspects of life science: **Small Crawling & Flying Animals; and Animal Growth & Changes**. In each section you will find teacher notes designed to provide you guidance with the learning intention, the success criteria, materials needed, a lesson outline, as well as provide some insight on what results to expect when the experiments are conducted. Suggestions for differentiation are also included so that all students can be successful in the learning environment.

Tracy Bellaire is an experienced teacher who continues to be involved in various levels of education in her role as Differentiated Learning Resource Teacher in an elementary school in Ontario. She enjoys creating educational materials for all types of learners, and providing tools for teachers to further develop their skill set in the classroom. She hopes that these lessons help all to discover their love of science!

Published in Canada by:
On The Mark Press
Belleville, ON
www.onthemarkpress.com

Funded by the
Government
of Canada

OTM2161   ISBN: 9781487710231
© On The Mark Press

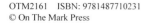

At A Glance

Learning Intentions

	Invertebrates	A Study of Invertebrates	Birds	Fish	Reptiles and Amphibians	Mammals	Life Cycles	Human Development	Co-Existing	An Aboriginal Connection
Knowledge and Understanding Content										
Recognize different types of invertebrates and describe their physical characteristics	•									
Research and describe the behavioral characteristics of different types of invertebrates		•								
Recognize the physical characteristics of birds, make comparisons between birds in flight, and research to learn more about bird behavior			•							
Recognize the physical characteristics of fish, and research to learn more about fish behavior				•						
Recognize the physical characteristics of reptiles and amphibians, and research to learn more about a reptile or an amphibian					•					
Identify different categories of mammals and their physical characteristics; research to learn more about a mammal						•				
Identify and describe the life cycle stages of different animals							•			
Describe the life cycle stages of a human, and determine the types of food that is needed for growth and development								•		
Explain how animals and humans co-exist on Earth and describe how they depend on each other for survival									•	
Research and describe the importance of certain animals to the lives of Aboriginal people										•
Thinking Skills and Investigation Process										
Make predictions, formulate questions, and plan an investigation				•						•
Gather and record observations and findings using drawings, tables, written descriptions	•	•	•	•	•	•	•	•	•	•
Recognize and apply safety procedures in the classroom	•	•	•	•	•	•	•	•	•	•
Communication										
Communicate the procedure and conclusions of investigations using demonstrations, drawings, and oral or written descriptions, with use of science and technology vocabulary	•	•	•	•	•	•	•	•	•	•
Application of Knowledge and Skills to Society and the Environment										
Identify the impacts that animals and humans have on each other, and recognize ways negative impacts could be minimized									•	
Identify the positive impact that human activity has on animals and their habitat, and recognize ways this positive impact could be jeopardized									•	

OTM2161   ISBN: 9781487710231
© On The Mark Press

TABLE OF CONTENTS

OTM2161   ISBN: 9781487710231
© On The Mark Press

Teacher Assessment Rubric

Student's Name: _____   Date: _____

Success Criteria	Level 1	Level 2	Level 3	Level 4
Knowledge and Understanding Content				
Demonstrate an understanding of the concepts, ideas, terminology definitions, procedures and the safe use of equipment and materials	Demonstrates limited knowledge and understanding of the content	Demonstrates some knowledge and understanding of the content	Demonstrates considerable knowledge and understanding of the content	Demonstrates thorough knowledge and understanding of the content
Thinking Skills and Investigation Process				
Develop hypothesis, formulate questions, select strategies, plan an investigation	Uses planning and critical thinking skills with limited effectiveness	Uses planning and critical thinking skills with some effectiveness	Uses planning and critical thinking skills with considerable effectiveness	Uses planning and critical thinking skills with a high degree of effectiveness
Gather and record data, and make observations, using safety equipment	Uses investigative processing skills with limited effectiveness	Uses investigative processing skills with some effectiveness	Uses investigative processing skills with considerable effectiveness	Uses investigative processing skills with a high degree of effectiveness
Communication				
Organize and communicate ideas and information in oral, visual, and/or written forms	Organizes and communicates ideas and information with limited effectiveness	Organizes and communicates ideas and information with some effectiveness	Organizes and communicates ideas and information with considerable effectiveness	Organizes and communicates ideas and information with a high degree of effectiveness
Use science and technology vocabulary in the communication of ideas and information	Uses vocabulary and terminology with limited effectiveness	Uses vocabulary and terminology with some effectiveness	Uses vocabulary and terminology with considerable effectiveness	Uses vocabulary and terminology with a high degree of effectiveness
Application of Knowledge and Skills to Society and Environment				
Apply knowledge and skills to make connections between science and technology to society and the environment	Makes connections with limited effectiveness	Makes connections with some effectiveness	Makes connections with considerable effectiveness	Makes connections with a high degree of effectiveness
Propose action plans to address problems relating to science and technology, society, and environment	Proposes action plans with limited effectiveness	Proposes action plans with some effectiveness	Proposes action plans with considerable effectiveness	Proposes action plans with a high degree of effectiveness

4

OTM2161   ISBN: 9781487710231
© On The Mark Press

Student Self Assessment Rubric

Name: _____ Date: _____

Put a check mark ✔ in the box that best describes you:

	Always	Almost Always	Sometimes	Needs Improvement
I am a good listener.		✓		
I followed the directions.		✓		
I stayed on task and finished on time.		✓		
I remembered safety.	✓			
My writing is neat.	✓			
My pictures are neat and colored.	✓			
I reported the results of my experiment.				
I discussed the results of my experiment.				
I know what I am good at.	✓			
I know what I need to work on.	✓			

1. I liked _____

_____

2. I learned _____

_____

3. I want to learn more about _____

_____

INTRODUCTION

The activities in this book have two intentions: to teach concepts related to life science and to provide students the opportunity to apply necessary skills needed for mastery of science and technology curriculum objectives.

Throughout the experiments, the scientific method is used. The scientific method is an investigative process which follows five steps to guide students to discover if evidence supports a hypothesis.

1. **Consider a question to investigate.**
   For each experiment, a question is provided for students to consider. For example, "How many times per minute does a goldfish breathe?"

2. **Predict what you think will happen.**
   A hypothesis is an educated guess about the answer to the question being investigated. For example, "I believe that a goldfish breathes about 20 times per minute". A group discussion is ideal at this point.

3. **Create a plan or procedure to investigate the hypothesis.**
   The plan will include a list of materials and a list of steps to follow. It forms the "experiment".

4. **Record all the observations of the investigation.**
   Results may be recorded in written, table, or picture form.

5. **Draw a conclusion.**
   Do the results support the hypothesis? Encourage students to share their conclusions with their classmates, or in a large group discussion format.

The experiments in this book fall under ten topics that relate to two aspects of life science: **Small Crawling and Flying Animals; and Animal Growth and Changes.** In each section you will find teacher notes designed to provide you guidance with the learning intention, the success criteria, materials needed, a lesson outline, as well as provide some insight on what results to expect when the experiments are conducted. Suggestions for differentiation are also included so that all students can be successful in the learning environment.

ASSESSMENT AND EVALUATION:

Students can complete the Student Self-Assessment Rubric in order to determine their own strengths and areas for improvement. Assessment can be determined by observation of student participation in the investigation process. The classroom teacher can refer to the Teacher Assessment Rubric and complete it for each student to determine if the success criteria outlined in the lesson plan has been achieved. Determining an overall level of success for evaluation purposes can be done by viewing each student's rubric to see what level of achievement predominantly appears throughout the rubric.

OTM2161   ISBN: 9781487710231
© On The Mark Press

INVERTEBRATES

LEARNING INTENTION:

Students will learn about the different types of invertebrates and their physical characteristics.

SUCCESS CRITERIA:

- identify the body parts of an insect, a spider, a worm, and a crustacean
- list types of common insects
- research and retell some facts on spiders
- create a home for earthworms
- record observations about earthworm activity
- create a home for a hermit crab
- describe the needs of hermit crabs in order to live
- describe the growth and changes in a hermit crab over time

MATERIALS NEEDED:

- a copy of "What's an Invertebrate? – Part One" worksheet 1, 2, 3, and 4 for each student
- a copy of "What's an Invertebrate? – Part Two" worksheet 5 and 6 for each student
- a copy of "What's an Invertebrate? – Part Three" worksheet 7 for each student
- a copy of "A Home for Earthworms" worksheet 8 and 9 for each student
- a copy of "What's an Invertebrate? – Part Four" worksheet 10 for each student
- a copy of "A Home for a Hermit Crab" worksheet 11 and 12 for each student
- access to the internet or local library
- soil such as sand and loam or topsoil (enough to fill large jars for each student)
- a hammer and a nail, masking tape, a jug of water, a few small cups
- earthworms (2 or 3 per student)
- vegetable or fruit scraps

- large sheets of black construction paper (2 per student)
- chart paper, markers, pencil crayons, clipboards, pencils
- *painting paper, paint, paint brushes, modeling clay (optional materials)*
- *visit a local pet store or go on line at www.petsmart.com to gather the materials and information needed in order to create a home for a hermit crab

PROCEDURE:

*This lesson can be done as one long lesson, or be divided into six shorter lessons.

1. Using worksheets 1 and 2, do a shared reading activity with the students. This will allow for reading practice and breaking down word parts to read the larger words. Along with the content, discussion of vocabulary words would be of benefit for their comprehension.

   Some interesting vocabulary words to focus on are:

   - invertebrate
   - antennae
   - complex eyes
   - thorax
   - image
   - insect
   - mandibles
   - head
   - facets
   - exoskeleton
   - organs
   - simple eyes
   - abdomen

2. Give students worksheets 3 and 4. They will list 5 insects that they know, and then compare lists with a classmate. Students *may* need to access the internet or visit a local library to find out more about the physical appearance of the insect they choose to draw and label, and to find out what it looked like upon hatching.

OTM2161   ISBN: 9781487710231
© On The Mark Press

3. Using worksheets 5 and 6, do a shared reading activity with the students. This will allow for reading practice and breaking down word parts to read the larger words. Along with the content, discussion of vocabulary words would be of benefit for their comprehension. Students will also need to access the internet or visit a library to find facts about spiders. An option is to come back as a large group and have students share their facts.

Some interesting vocabulary words to focus on are:

- arachnids
- poisonous glands
- vibrations
- abdomen
- venom
- fangs
- silk
- cephalothorax
- spinnerets
- prey

4. Using worksheet 7, do a shared reading activity with the students. This will allow for reading practice and breaking down word parts to read the larger words. Along with the content, discussion of vocabulary words would be of benefit for their comprehension.

Some interesting vocabulary words to focus on are:

- slither
- digests
- vegetation
- crumbly
- burrows
- earthworm
- castings
- tunnels
- moist
- soil
- creature
- nutrients

5. Students will create a home for earthworms. Give them worksheets 8 and 9, and the materials needed. Read through the materials needed, and what to do sections on worksheet 8 with the students to ensure their understanding of the task. Students will make and record observations of the earthworms home as it is created and 48 hours later. They will make a conclusion about the purpose earthworms have in creating nutrient rich soil.

6. Using worksheet 10, do a shared reading activity with the students. This will allow for reading practice and breaking down word parts to read the larger words. Along with the content, discussion of vocabulary words would be of benefit for their comprehension.

Some interesting vocabulary words to focus on are:

- crustaceans
- environment
- thorax
- protects
- head
- exoskeleton
- antennae
- abdomen
- jointed
- dangers

7. Together as a class, create a home for a hermit crab that can be maintained in the classroom. You will need to visit a local pet store, or go online at www.petsmart.com to gather the information on the materials you will need and how to care for it daily.

8. Once the hermit crab's home is created, give students worksheet 11. They will illustrate what it looks like.

9. After a few weeks or months, give students worksheet 12. They will detail how to care for the hermit crab and comment on any changes they have noticed.

DIFFERENTIATION:

Slower learners may benefit by working in a small group with teacher support to orally answer the questions on worksheet 12. Responses could be recorded on one large chart paper, and then displayed in the classroom or on a bulletin board. This chart paper could then be surrounded by an art activity (painting) done by students that depicts the hermit crab in its environment.

For enrichment, faster learners could use modeling clay to create three dimensional models of the four types of invertebrates (insect, spider, worm, crustacean). These could be left on a display table.

OTM2161   ISBN: 9781487710231
© On The Mark Press

What is an Invertebrate? – Part One

An **invertebrate** is an animal with no backbone. Some invertebrates like **insects** have an **exoskeleton**, which is a skeleton that is on the outside of their bodies.

Insects have bodies that are made up of three parts. These parts are the **head**, the **thorax**, and the **abdomen**. They have three pairs of legs that grow out from the front, middle, and back of the thorax. The head of an insect has two **antennae**.

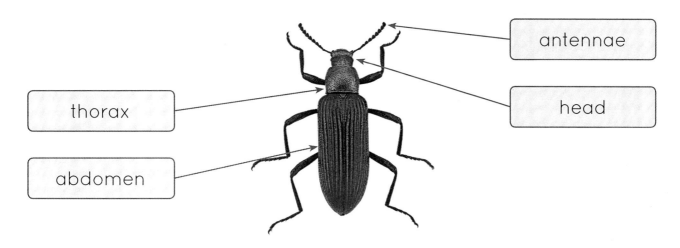

antennae

head

thorax

abdomen

Winged insects like the dragonfly, have four wings that also grow out from the thorax.

Did you know that the antennae on some insects, like this red ant, have organs for smell as well as touch?

OTM2161   ISBN: 9781487710231
© On The Mark Press

Did you know that insects can have a pair of **complex eyes** and **simple eyes** too? Let's learn more about this!

Complex eyes have many small lenses called **facets**, and they can see colour. The parts that each facet sees, comes together to make one sharp image. So, insects see very well with their complex eyes.

A housefly has about 4000 lenses inside of each of its compound eyes.

If an insect has simple eyes too, they will be found in a form of three on top of its head. These eyes are simple because they can only see light and dark, not colour.

A form of 3 simple eyes, sit on the top of this paper wasp's head.

FAST FACT!

Many insects have mandibles, which are a pair of jaws. The jaws of an insect close sideways, unlike our own which close up and down.

OTM2161   ISBN: 9781487710231
© On The Mark Press

1. List 5 insects that you know:

- Butterfly
- fly
- Laddy bug
- ant
- Beetle

2. Compare your list with another classmate's list. Name 2 insects from their list that are not on yours.

- fruit fly
- Potato bug.

DID YOU KNOW?

All invertebrates lay eggs. That means all the insects on your list are egg laying creatures!

For some insects, when they hatch out of the egg, they look like their adults. For example, when an ant is hatched, it looks like a small ant.

For some insects, when they hatch out of the egg, they go through changes before they look like their adults. For example, when a caterpillar is hatched, it will become a pupa before it turns into an adult butterfly.

Choose one insect from your list.

1. Draw it in the box below.

2. Label its **antennae**, **head**, **thorax**, **abdomen**, **legs**, **eyes**, **mandibles**.

3. Label its **wings** if it has some.

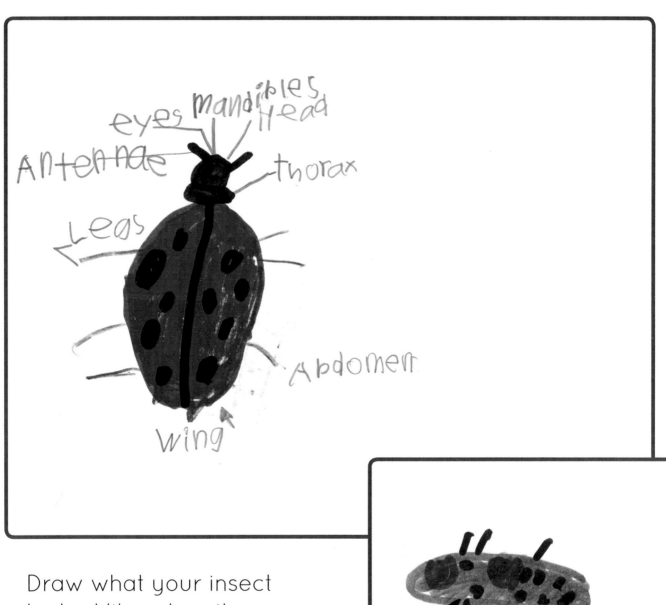

Draw what your insect looked like when it hatched from its egg.

OTM2161   ISBN: 9781487710231
© On The Mark Press

What is an Invertebrate? – Part Two

Another group of invertebrates are **arachnids**. Arachnids are creatures that have bodies made up of two parts. They have eight legs. They do not have wings or antennae. Can you guess what creature is part of this group? It is the spider!

A spider, like all other invertebrates, does not have a backbone. It has an **exoskeleton**. As the spider grows, it sheds its first exoskeleton and then stretches out as a new, larger exoskeleton forms.

The front part of a spider is called the **cephalothorax**.

The cephalothorax has the spider's eyes, brain, mouth fangs, and stomach. A spider's eight legs grow out of this part of its body.

If a spider was poisonous, this is where its poisonous glands would be.

The back part of the spider is called the **abdomen**.

A spider's **spinnerets** are at the back end of the abdomen. A spider makes **silk** in its spinnerets. Some spiders use this silk to spin webs.

Did you know that a spider's body has oil on it that helps it from sticking to its own web?

OTM2161   ISBN: 9781487710231
© On The Mark Press

SPINNING OUT THE FACTS!

Did you know that a spider has 48 knees? Each one of its 8 legs has 6 joints on it! At the end of each of its legs, a spider has small claws.

There are many hairs that cover a spider's legs. These hairs can sense vibrations or movements. They can also pick up smells in the air.

Some spiders have venom in their fangs. They use the venom in their fangs to catch their prey. The shot of venom stills their prey, and then the spider goes in for a tasty snack!

Some spiders have poisonous venom. If a spider's prey gets poked by its fangs, the poison will turn its guts to liquid. Then, the spider will suck it out!

Now it is your turn to spin out the facts! Use the internet or visit a local library to learn about spiders. Tell two facts about spiders.

1. They are arachnids.

2. Some tarantulas can live for more then 20 years

OTM2161 ISBN: 9781487710231
© On The Mark Press

What is an Invertebrate? – Part Three

Worms are also invertebrates. They have no legs, so they slither or inch their bodies along to move from place to place. There are different kinds of worms, but the one that is most common is the earthworm. Let's learn more about this creature!

Where Do They Live?

Earthworms make their homes in soil. Some earthworms live in the roots of grass and some live in vegetable or flower gardens. Some earthworms will even make their burrows under leaves around tree roots.

The earthworm's burrow is long and dark. Earthworms like to stay in their burrows because it is dark and moist.

They come out at night to eat leaves and grass, or other vegetation.

What Do They Do?

The earthworm is like "nature's plow" because their burrowing of tunnels brings air to the roots of plants. They make the earth crumbly so water can get into the soil.

Worms help water flow through the soil!

When an earthworm eats and then digests its food, its castings add nutrients to the soil that plants need to grow.

OTM2161   ISBN: 9781487710231
© On The Mark Press

Name:

A Home for Earthworms

Some people make worm farms to make rich soil for their gardens at home. Let's give this a try!

You'll need:

- a large jar with a lid
- a hammer and a nail
- masking tape
- 2 or 3 earthworms
- 2 sheets of black construction paper
- soil (loam and sand)
- vegetable or fruit scraps
- a small cup of water

What to do:

1. Fill the jar three-quarters full of loam soil. Add a layer of sand on top.

2. Put some vegetable or fruit scraps on top of the sand.

3. Moisten the soil with a bit of water. Then add the earthworms.

4. Using the hammer and a nail, **your teacher** will make some holes in the lid of the jar.

5. Put the lid on the jar to close it.

6. On worksheet 4, draw what you see in the jar.

7. Cover the sides of the jar with black construction paper so that no light can get in. **Earthworms like it dark!**

8. Leave the jar for 2 days.

9. After 2 days, take off the paper. Record what you see now on worksheet 9.

OTM2161   ISBN: 9781487710231
© On The Mark Press

Let's Observe!

Draw your observations of what is in the jar.

This is what it looked like in the jar before it was covered up:

This is what it looked like in the jar after it was left for 2 days:

Let's Conclude

What did the earthworms do?

_____

_____

_____

OTM2161   ISBN: 9781487710231
© On The Mark Press

What is an Invertebrate? – Part Four

Another group of invertebrates are **crustaceans**. They live mostly in water. Some examples of crustaceans are lobsters, crabs, and shrimp.

lobster

crab

shrimp

Crustaceans have many legs with jointed parts. Some crustaceans like shrimp and crabs can swim, but others like lobsters must use their legs to move from place to place.

The body of a crustacean is made up of the head, the thorax, and the abdomen. Its eyes are in the middle of its head. Did you know that crustaceans have two pairs of antennae? They use their antennae to sense what is in their environment, like food, or danger.

DID YOU KNOW?

A crustacean's exoskeleton protects its body. Its body will grow, but its exoskeleton will not. So, it will shed its shell when its body gets too big for it, and grow another one. Except for the hermit crab! The hermit crab cannot make his own shell, so it hides in shells left behind by other crustaceans.

OTM2161 ISBN: 9781487710231
© On The Mark Press

Name:

A Home for a Hermit Crab

Work together with your teacher and classmates to set up a home for a hermit crab. Watch for changes that happen as the hermit crab grows. Record what you see.

This is the home that we made for the hermit crab:

OTM2161   ISBN: 9781487710231

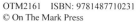

Let's Observe!

What do you do to care for the hermit crab so that its needs are met?

1. _____

_____

_____

_____

2. _____

_____

_____

_____

3. _____

_____

_____

_____

What growth and changes have you noticed in the hermit crab?

_____

_____

_____

_____

OTM2161   ISBN: 9781487710231
© On The Mark Press

A STUDY OF INVERTEBRATES

LEARNING INTENTION:

Students will learn about the behavioral characteristics of invertebrates, through research and through compare and contrast.

SUCCESS CRITERIA:

- choose an invertebrate to research
- describe its appearance, where it lives, what it eats, its predators, its day and night time activity, and special changes it experiences
- compare and contrast two invertebrates
- display all information using pictures, diagrams, and written descriptions
- orally share one interesting fact about an invertebrate

MATERIALS NEEDED:

- a copy of "Investigating an Invertebrate!" worksheet 1, 2, 3, and 4 for each student
- a copy of "The Invertebrate Exchange!" worksheet 5 and 6 for each student
- access to the internet or local library
- Bristol board (1/2 piece per student)
- chart paper, markers, glue sticks, pencils, clipboards, pencil crayons
- light colored construction paper, scissors *(optional materials for trivia card creation)*

PROCEDURE:

*This lesson can be done as one long lesson, or done in two or three shorter lessons.

1. Engage students in a brainstorming activity to name different types of invertebrates that they know. Record students' ideas on chart paper.

2. Instruct students to choose an invertebrate that they would like to learn more about. Give students worksheets 1, 2, 3, and 4, a clipboard and a pencil. With access to the internet or resources at a local library, students will investigate and record information about their chosen invertebrate. Upon completion, give each student a half piece of large Bristol board to display their work on.

3. Divide students into pairs (ensuring their chosen invertebrates are different species). Give students worksheets 5 and 6, and a clipboard and pencil. They will compare and contrast their invertebrates. This will allow for further learning about other species of invertebrates.

4. Invite students to participate in a "Did you know?" activity. They will come back to the large group, with one interesting fact that they would like to share with the rest of the class. Instruct them to form their sentence about their interesting fact as "Did you know...?" This activity will promote rich discussion and use of life science terminology about invertebrates.

5. Display Bristol boards containing students' work around the school, or in the classroom.

DIFFERENTIATION:

Slower learners may benefit by working as a small group with teacher support and direction to complete worksheets 1, 2, 3, and 4. One invertebrate should be chosen, and then to lessen the work load, each student in the group could be assigned a section to research, and then share the findings with the small group in order to complete the worksheets.

For enrichment, faster learners could create a trivia card game using the facts that they learned from the "Did you know...?" activity. Trivia questions should begin with "What invertebrate...?"

OTM2161   ISBN: 9781487710231
© On The Mark Press

Investigating an Invertebrate!

You have learned about different types of invertebrates. Now it is time to choose one to learn more about what it looks like, where it lives, what it eats, and how it lives. Let's get started!

The invertebrate I am investigating is _Garden snail_ .

This is a diagram of what it looks like:

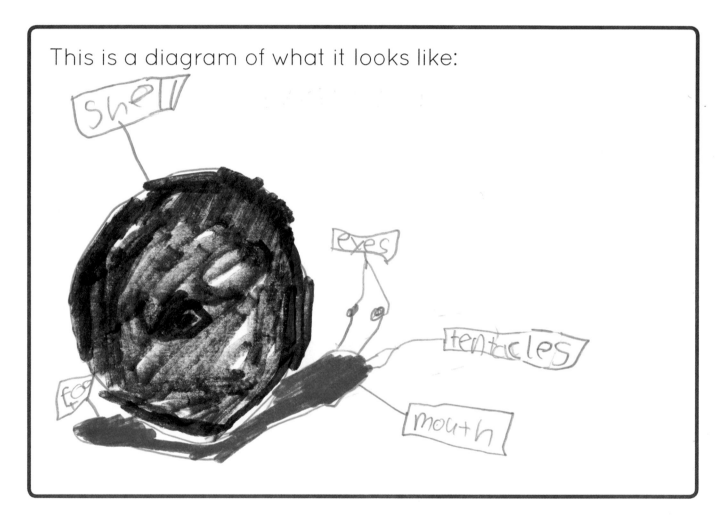

Where does it live?

Native to Europe but now it lives here.

OTM2161 ISBN: 9781487710231
© On The Mark Press

Make a list of things that your invertebrate eats.

- fruit trees
- herbs
- flowers
- tree bark
-

How does it get its food?

They have a toothed rippon that breaks up its food.

Draw and label some of its predators.

frog

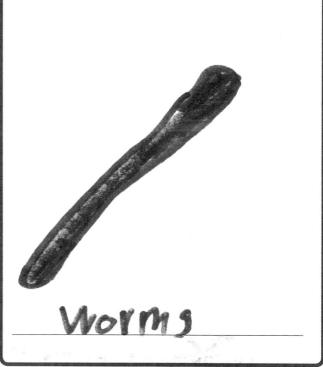

Worms

Complete the sentences and add drawings.

In the daytime, my invertebrate _Sleeps, unless it is_

wet out.

_____

At night time, my invertebrate _eats._

_____

_____

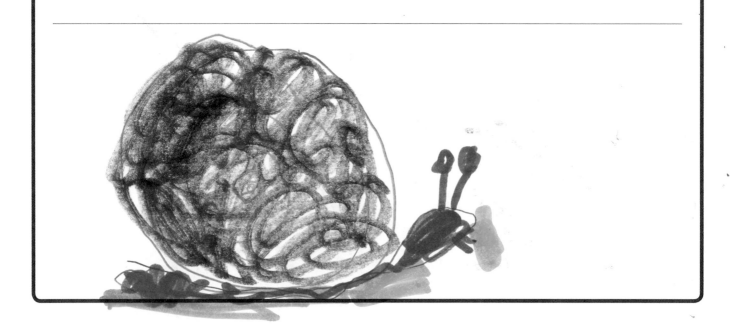

OTM2161   ISBN: 9781487710231
© On The Mark Press

Name:

Tell about any special changes that happen to your invertebrate. For example:

- Does it migrate?

- Does it change its form?

- Does it molt?

_____

_____

_____

_____

_____

_____

Wrapping Up!

By investigating this invertebrate, I learned...

_____

_____

_____

_____

I still wonder about...

_____

_____

_____

_____

_____

OTM2161   ISBN: 9781487710231
© On The Mark Press

Name:

The Invertebrate Exchange!

Partner up with a classmate. Exchange information about the invertebrates that you have investigated. Be sure that your classmate's invertebrate is different from yours!

This is the invertebrate that my partner investigated:

The _____

Do your invertebrates look the same in any way? Explain.

_____

_____

_____

Do your invertebrates look different in any way? Explain.

_____

_____

_____

OTM2161   ISBN: 9781487710231

Name:

This is what is on the menu for my partner's invertebrate:

- ◆ _____

- ◆ _____

- ◆ _____

- ◆ _____

Circle any food items that are the same for your invertebrate.

Talk with your partner about the day and night time activities of your invertebrates. Then complete the sentences below.

Our invertebrates both like to _____

_____

_____

My invertebrate likes to _____

_____

_____, but my partner's invertebrate likes to

_____

_____

OTM2161 ISBN: 9781487710231
© On The Mark Press

BIRDS

LEARNING INTENTION:

Students will learn about the physical characteristics of birds, make comparisons between birds in flight, and research to learn more about bird behavior.

SUCCESS CRITERIA:

- recognize the physical characteristics of different birds
- make observations of different birds in flight and record results in a chart
- choosing a bird to research, describe changes in its appearance as it grows
- describe where it lives, what it eats, how it moves, how it cares for its young
- display all information using pictures and written descriptions
- research and orally share one interesting fact about a bird

MATERIALS NEEDED:

- a copy of "Birds" worksheet 1, 2, and 3 for each student
- a copy of "On a Bird Watch" worksheet 4 and 5 for each student
- a copy of "Going on a Bird Study!" worksheet 6, 7, 8, and 9 for each student
- a stopwatch and a clipboard for each student
- access to the internet or local library, pencils, pencil crayons
- sheets of art paper, paint, and paint brushes (optional materials)

PROCEDURE:

*This lesson can be done as one long lesson, or be divided into four shorter lessons.**

1. Using worksheets 1, 2, and 3, do a shared reading activity with the students. This will allow for reading practice and breaking down of the larger words. Along with content, discussion of some vocabulary would be beneficial for comprehension.

Some interesting vocabulary words to focus on are:

- climates
- skeleton
- hatch
- beaks
- birds of prey
- migrate
- hollow
- talons
- scoop
- clawed feet
- vertebrate
- glide
- adapted
- swallow
- webbed feet

2. Explain to students that they will go on a bird watching exploration in order to spot, time, and compare the flight capabilities of different birds in the neighborhood. Give students worksheets 4 and 5, and a clipboard and pencil. After spending some time bird watching and recording flight results of different birds, students will work with a partner to discuss any interesting findings about the flight potential of different birds.

3. Give students worksheet 6. They will list 5 insects that they know, and then compare their list with a classmate's. They will choose one bird from their list to study.

4. Give students worksheets 7, 8, and 9. With access to the internet or by visiting a local library, students will find out more about the physical appearance of the bird they choose and what it looked like upon hatching. They will also need to find out what it eats, the ways it moves, how it cares for its young, and add an interesting fact of their own.

5. A follow up option is to have students showcase their picture of their bird to the large group and orally share the information on their 'bird fact – did you know?' card.

DIFFERENTIATION:

Slower learners may benefit by working as a small group with teacher support and direction to complete worksheets 7, 8, and 9. Choosing one bird, assign a section to each student in the group. Then they can share their findings with the small group in order to complete the worksheets. Omission of the fact card on worksheet 9 is an additional accommodation.

For enrichment, faster learners could paint a picture of their bird in its environment.

OTM2161   ISBN: 9781487710231
© On The Mark Press

Birds

Did you know that there are about 10 000 different kinds of birds in the world? They live all over the world, on almost every piece of land and in all climates. Some birds **migrate** from place to place. Other birds live in the same place year round.

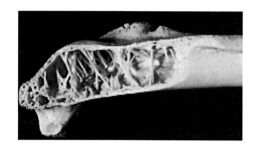

Birds are **vertebrate** animals. This means they have a skeleton on the inside of their body.

The inside of their bones are filled with tiny air sacs and their bones are hollow too. This makes them light enough to fly.

The feathers on birds keep them warm and they also help them to fly. The feathers form the shape of their wings. They can fold their wings in or hold them out.

Birds get power and lift from beating their wings. Some birds can glide in the air without flapping their wings for long distances.

FAST FACT!

Some birds like the ostrich, do not fly. Instead they use their strong legs to run fast. Penguins, which live in the Antarctic, cannot fly either. Instead they use their wings as flippers and swim fast under water.

OTM2161   ISBN: 9781487710231
© On The Mark Press

Name:

All birds lay eggs. A young bird grows inside the egg until it is ready to **hatch**. Eggshells are very hard in large birds like the ostrich, but are very soft in small birds such as the robin.

This mother robin has caught a worm to feed to her babies in the nest.

Young birds need a lot of care from their parents. The parent bird will catch insects and other food to feed it to their babies.

When baby birds are older and have grown feathers and strong wings, the parents teach them to fly. Then they are ready to catch their own food.

While some birds eat things such as insects or worms, other birds eat small animals like mice, snakes, and even fish too! These birds are called **birds of prey**. They have **talons**, which are clawed feet. Their feet are made for grabbing prey.

Other birds like loons and ducks, live on water. They eat insects and fish. These birds have **webbed feet** to help them swim.

OTM2161   ISBN: 9781487710231
© On The Mark Press

Name:

A bird's mouth is **adapted** to what it eats. For example, birds of prey have sharp hooked **beaks** to catch and eat their prey. Did you know that the beak of an eagle is made of keratin? It is always growing, just like the fingernails of a human!

Some birds, like the swallow, have large mouths but small beaks. These birds can catch insects while flying with their mouths open!

Some birds have sharp beaks adapted for biting seeds and cracking shells, like this parrot.

Some birds, like the pelican have large mouths. They scoop up small fish, and swallow them whole.

OTM2161   ISBN: 9781487710231
© On The Mark Press

On a Bird Watch

Do you know how long it takes a bird to go from land to air? Well, each bird is different. Do a bird study to find out more. Let's get into nature!

You'll need:

- a copy of "On a Bird Watch" worksheet

- a clipboard

- a pencil

- a stopwatch

What to do:

1. Outside, look around for birds that are ready to fly into the air. When you see one, watch it carefully. When you see a bird taking off, hit the "start" button on the stopwatch. Stop timing when the bird gets into the sky.

2. On worksheet 5, record the kind of bird and the time that the bird took to get into the air.

3. Repeat steps 1 and 2, looking for 2 other birds.

4. Compare your notes with a partner. What do you notice?

OTM2161   ISBN: 9781487710231
© On The Mark Press

Name:

Let's Observe

Birds	Time in Flight	Notes About This Bird
Bird #1 Bule Jay		
Bird #2 Sparrow		
Bird #3 Cardinal		

Compare notes with a partner. What is interesting?

_____

_____

_____

_____

_____

Going on a Bird Study!

1. List 5 birds that you know:

   - Chickadee
   - Loon
   - Duck
   - Blue Jay
   - Woodpecker
   - Wren

2. Compare your list with a classmate's list. Name 2 birds from their list that are not on yours.

   - Hawk
   - Crow
   -
   -

3. Circle **one** bird from the lists above that you want to learn more about.

   What do you wonder about this bird?

   What dose it eat?
   What colour are it's eggs?
   What dose it build it's nest with?

OTM2161   ISBN: 9781487710231
© On The Mark Press

In the box, draw a picture of your bird that shows where it lives.

Draw what your bird looked like when it hatched as a baby from its egg.

How has it changed from when it hatched to now?

It turns brown.
It feeds it self.
Ii gets bigger.

Name:

Draw pictures of some things that your bird likes to eat.
Label them.

spiders

flys

My bird's diet

Beetel

Lady bug

Circle all the ways that your bird moves.

fly   swim   walk   run

OTM2161   ISBN: 9781487710231
© On The Mark Press

Name:

Tell about some things that your bird does to care for its babies.

◆ _____

_____

_____

◆ _____

_____

_____

◆ _____

_____

_____

Make a fact card about your bird.

DID YOU KNOW?

_____

_____

_____

_____

_____

_____

_____

OTM2161   ISBN: 9781487710231
© On The Mark Press

FISH

LEARNING INTENTION:

Students will learn about the physical characteristics of fish, and research to learn more about fish behavior.

SUCCESS CRITERIA:

- make observations of the breathing rate of a goldfish
- record results using a chart and a diagram
- make conclusions about the breathing rate of a goldfish
- label the main parts of a fish
- identify and label some fresh water and some salt water fish
- research and record some facts about fish

MATERIALS NEEDED:

- a copy of "Fish" worksheet 1 for each student
- a copy of "It's All in the Gills!" worksheet 2 and 3 for each student
- a copy of "A Diagram of a Fish" worksheet 4 for each student
- a copy of "Let's Go Fresh Water Fishing" worksheet 5 for each student
- a copy of "Let's Go Salt Water Fishing" worksheet 6 for each student
- a copy of "Fishing for the Facts! Worksheet 7 for each student
- a goldfish, a timer, a fish bowl (prepared according to instructions from a professional at a pet store)
- access to the internet or local library
- pencils, pencil crayons, clipboards

PROCEDURE:

*This lesson can be done as one long lesson, or be divided into four shorter lessons.

1. Using worksheet 1, do a shared reading activity with the students. This will allow for reading practice and breaking down of the larger words. Along with content, discussion of some vocabulary would be beneficial for comprehension.

   Some interesting vocabulary words to focus on are:
   - vertebrate
   - skeleton
   - oxygen
   - operculum
   - scales
   - fins
   - gills
   - carbon dioxide
   - protective
   - tail
   - blood vessels

2. Students will conduct an observational experiment to determine the breathing rate of a goldfish. *Depending on the size of the class, this activity could be done as a large group, or in small groups, with teacher direction and support. Give students worksheets 2 and 3, and the materials to conduct the observation. Students will make and record their prediction, observe the breathing rate of the goldfish, record responses, and make a conclusion about the results.

   (Students should conclude that the fish's mouth and operculum open and close an equal number of times. The fish opens its mouth to inhale water. Then the mouth closes and the operculum opens. The water that came through its mouth, passes over the gills and out the operculum.)

OTM2161   ISBN: 9781487710231
© On The Mark Press

3. Give students worksheet 4. They will label the main parts of a fish. Students may need to access the local library or internet to assist them in completing the diagram.

4. Give students worksheets 5 and 6. They will match the names of each fish to its picture. Students may need to access the local library or internet to find pictures of the different types of fish on the worksheet so that they can match their pictures to a name label.

5. Give students worksheet 7. With access to the local library or internet, students will research and record some facts about fish.

DIFFERENTIATION:

Slower learners may benefit by working as a small group with teacher support and direction to complete worksheet 4. Omission of worksheet 7 is an additional accommodation.

For enrichment, faster learners could choose a fish from worksheet 6 or 7, and then research 3 interesting facts about it. These facts could be shared orally with the large group as a follow up option.

OTM2161   ISBN: 9781487710231
© On The Mark Press

Fish

Did you know that there are about 28 000 different kinds of fish in the world? Fish can be found in lakes, rivers, oceans, streams, or ponds. Some live in fresh water and some live in salt water.

Fish are vertebrates. They have a skeleton inside their bodies. The outside of their bodies is covered in scales. They have fins and a tail which help them to swim in the water. Some fish have more fins than others.

Did You Know?

Most fish get their oxygen from water and not from air. Fish get oxygen from water through tiny blood vessels that are in their gills. The gills are under a protective flap called the

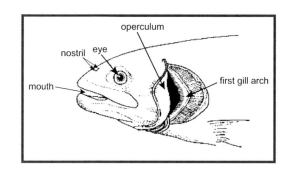

operculum. There is a set of gills on each side of the fish's body.

A fish opens its mouth to take in water. When its mouth is open, the operculum is closed. Then its mouth will close and the operculum will open.

The water that the fish took in its mouth passes over its gills and through its blood vessels. The fish uses the oxygen that is in the water to breathe. As water goes out the operculum, the fish breathes out carbon dioxide.

OTM2161   ISBN: 9781487710231
© On The Mark Press

It's All in the Gills!

Now that you know how a fish breathes under water, let's watch it happen!

> **Question**: How many times per minute does a goldfish breathe?

Materials Needed:

• a gold fish

• a fish bowl

• a timer

What To Do:

1. Make a prediction about the answer to the question then record it on worksheet 3.

2. Locate the fish's mouth.

3. Set the timer for one minute.

4. During the minute, count the number of times the goldfish opens and closes its mouth.

5. Record your answer on the chart on worksheet 3.

6. Repeat steps 3 – 5, this time watching the operculum open and close.

7. Draw a diagram to show your observations of the goldfish breathing in the water.

8. Make a conclusion about what you have observed.

OTM2161   ISBN: 9781487710231
© On The Mark Press

Let's Predict

How many times per minute does a goldfish breathe?

_____

_____

Let's Observe

Record the number of times the goldfish opened and closed its mouth in one minute.	Record the number of times the goldfish opened and closed its operculum in one minute.
_____ times in one minute	_____ times in one minute
Diagram of the goldfish as it breathes in the water:	

Let's Conclude

Explain the results. Why did the goldfish open and close its mouth and its operculum about the same number of times?

_____

_____

OTM2161   ISBN: 9781487710231
© On The Mark Press

A Diagram of a Fish

Use the words in the Word Box below to label the parts of a fish.

eye	nostril	tail fin mouth
scales	anal fin	ventral fin gill
		soft dorsal fin spiny dorsal fin

mouth

nostril

eye

Scales

gill

spiny dor

tail fin

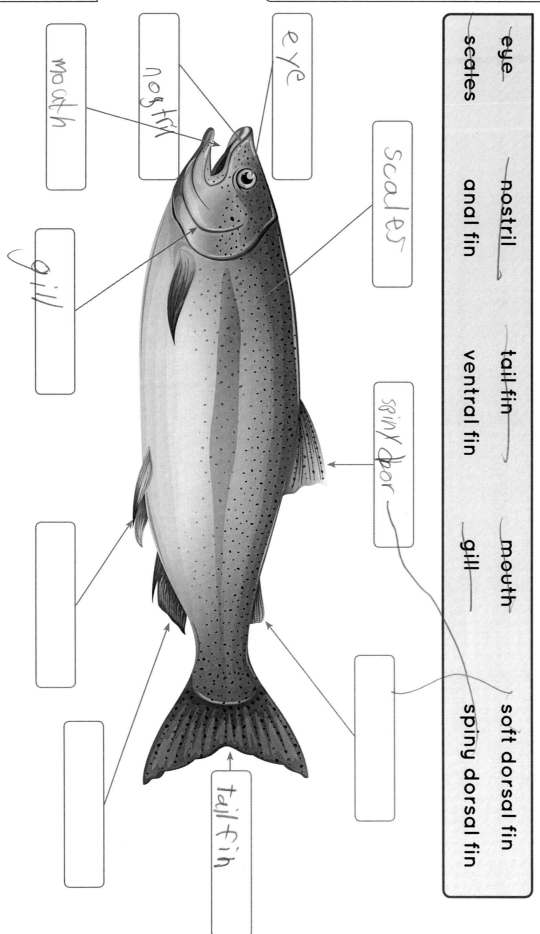

OTM2161   ISBN: 9781487710231
© On The Mark Press

Name:

Let's Go Fresh Water Fishing!

Draw a line from each fishing line to the correct fish.

walleye

northern pike

perch

large mouth

rainbow trout

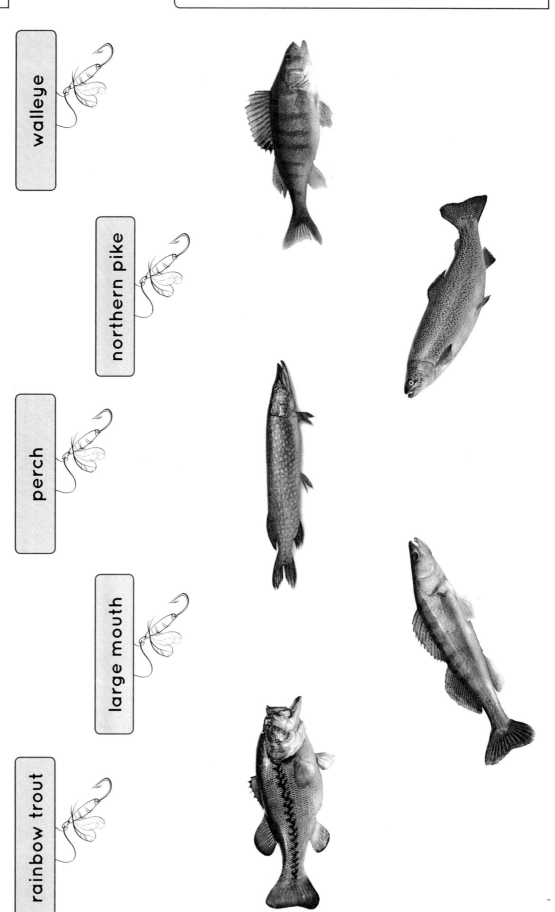

OTM2161   ISBN: 9781487710231
© On The Mark Press

Let's Go Salt Water Fishing!

Draw a line from each fishing line to the correct fish.

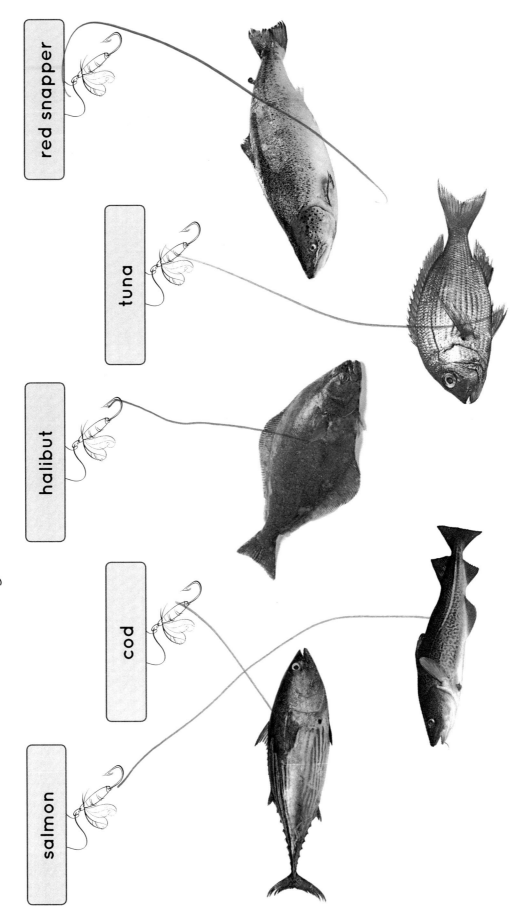

red snapper

tuna

halibut

cod

salmon

Fishing for the Facts!

Visit your local library or access the internet to find out some fish facts by answering the questions below.

Fish Facts!

Q. Why do some fish travel in schools?

A. _____

_____

_____

Q. Why do some fish have bright colors?

A. _____

_____

_____

Q. Do all fish lay eggs? Explain.

A. _____

_____

_____

Q. Can fish hear? Explain.

A. _____

_____

_____

OTM2161   ISBN: 9781487710231
© On The Mark Press

REPTILES AND AMPHIBIANS

LEARNING INTENTION:

Students will learn about the physical characteristics of reptiles and amphibians, and research to learn more about a reptile or an amphibian.

SUCCESS CRITERIA:

- recognize the physical characteristics of reptiles and amphibians
- identify reptiles and amphibians
- research two interesting facts about a chosen reptile or amphibian
- describe where it lives, and how it changes in its appearance as it grows
- display information using pictures and written descriptions

MATERIALS NEEDED:

- a copy of "Reading About Reptiles!" worksheet 1 for each student
- a copy of "About Those Amphibians!" worksheets 2 and 3 for each student
- a copy of "Reptile or Amphibian?" worksheet 4 for each student
- a copy of "Research It!" worksheets 5 and 6
- access to the internet or local library
- pencils, pencil crayons, clipboards

PROCEDURE:

*This lesson can be done as one long lesson, or done in three or four shorter lessons.

1. Using worksheets 1, 2, and 3, do a shared reading activity with the students. This will allow for reading practise and breaking down of the larger words. Along with content, discussion of some vocabulary would be beneficial for comprehension.

Some interesting vocabulary words to focus on are:

- tuatara
- nutrients
- cold
- gills
- amphibian
- herbivores
- temperature
- prey
- blooded
- carnivores
- hibernate
- reptile
- nocturnal
- climate
- polar
- control

2. Give students worksheet 4, and a red and a green pencil crayon. They will identify the reptiles by circling them in red and identify the amphibians by circling them in green. Students may need to access the internet to aid them in the identifications.

   (top row) – tortoise (reptile), gecko (reptile), newt (amphibian)
   (second row) – frog (amphibian), cobra (reptile), lizard (reptile)
   (third row) – iguana (reptile), salamander (amphibian), turtle (reptile)
   (bottom row) – toad (amphibian), crocodile (reptile), caecilian (amphibian)

3. Give students worksheets 5 and 6. With access to the internet or by visiting a local library, students will find out more about a reptile or an amphibian of their choice. Once completed, these worksheets could be displayed on a bulletin board or around the school.

DIFFERENTIATION:

Slower learners may benefit by working with a strong peer to complete worksheet 4. This would allow for discussion, and for guidance to locate the animals on the internet, if necessary. An additional accommodation is for these learners to research only one fact about their chosen animal, on worksheet 5.

For enrichment, faster learners could find out what their chosen reptile or amphibian likes to eat and how it catches its prey.

OTM2161   ISBN: 9781487710231
© On The Mark Press

Reading About Reptiles!

Did you know that there are over 8000 different kinds of reptiles in the world?

Reptiles are animals such as snakes, lizards, turtles, crocodiles, alligators, and tuataras. Tuataras live in New Zealand. They are nocturnal reptiles. They sleep in the day and hunt for food at night.

Tuataras prey upon insects, spiders, frogs, and other small reptiles.

Reptiles can be found all over the world, except in the polar areas because it is too cold there for them.

Reptiles are **vertebrate** animals. This means they have a skeleton on the inside of their body. Their skin is covered in scales, and some have a bony plate called a shell.

Reptiles are cold-blooded animals. This means that they cannot control their own body temperature. They need the heat from the sun to become warm and active. A reptile can get its warmth from the sun, heated rocks, logs, or soil.

FAST FACT!

Most reptiles lay eggs. When a baby hatches from its egg, it uses its egg tooth to break the shell open. Soon after, it loses its egg tooth.

OTM2161   ISBN: 9781487710231
© On The Mark Press

About Those Amphibians!

There are about 7000 different kinds of amphibians in the world. There are three groups of amphibians. The first group is tailless, like frogs and toads. The second group is tailed like salamanders and newts. The third group is called caecilians, these are worm-like creatures.

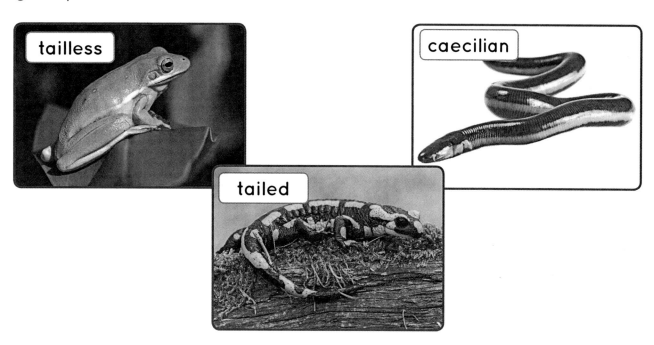

tailless

caecilian

tailed

Amphibians are cold-blooded animals. They need the heat from the sun to warm up their body temperature to help them get active. Because of this, amphibians that live in cooler climates hibernate in the winter.

Amphibians are **vertebrate** animals. This means they have a skeleton on the inside of their body. Most amphibians have smooth, moist skin. There are a few, like the toad, that have scaly skin.

Amphibians often shed parts of their skin and grow new skin. Some will eat their skin for the nutrients in it. Did you know that some amphibians, like the salamander, can even grow new arms and legs if they lose one?

OTM2161   ISBN: 9781487710231
© On The Mark Press

Amphibians live in places near water. They can be found in streams, forests, swamps, ponds, and lakes. Amphibians live in the water and on land. They begin their life in water with gills and tails. As they grow, their bodies change. They grow lungs and legs so that they can live on land too.

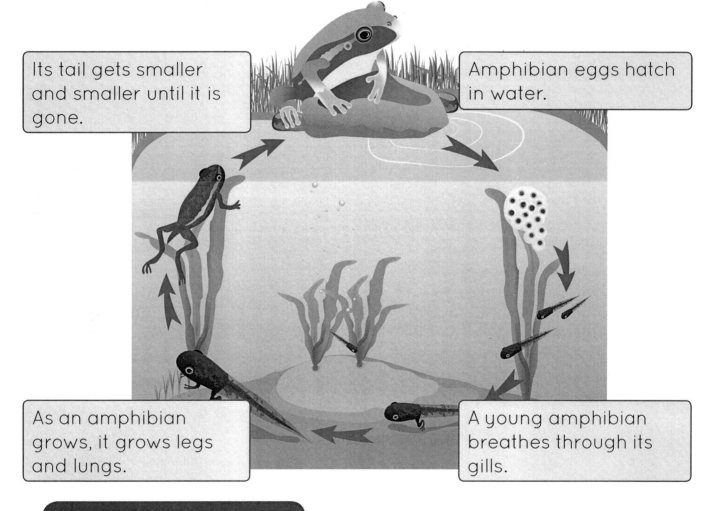

The adult amphibian lives on land and water.

Its tail gets smaller and smaller until it is gone.

Amphibian eggs hatch in water.

As an amphibian grows, it grows legs and lungs.

A young amphibian breathes through its gills.

FAST FACT!

Adult amphibians are mostly carnivores. They eat insects, slugs, and worms. Young amphibians are mostly herbivores. They eat small plants and algae.

OTM2161 ISBN: 9781487710231
© On The Mark Press

Reptile or Amphibian?

Use a **red** pencil crayon to circle the **reptiles**. Use a
green pencil crayon to circle the **amphibians**.

Name:

Research It!

Visit the library or access the internet to find out some interesting facts about a reptile or an amphibian that you would like to learn more about.

I am going to research about _____.

Is it a reptile or an amphibian _____?

Get the Facts!

Make two fact cards about the animal you chose.

DID YOU KNOW?

_____

_____

_____

_____

DID YOU KNOW?

_____

_____

_____

_____

Share your fact cards with a classmate!

OTM2161   ISBN: 9781487710231
© On The Mark Press

Name:

In the box, draw a picture of your animal that shows where it lives.

Draw what your animal looked like when it was born.

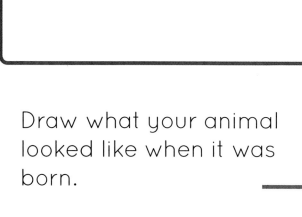

How has it has changed from when it was born to now?

_____

_____

_____

_____

OTM2161 ISBN: 9781487710231

MAMMALS

LEARNING INTENTION:

Students will learn about the different categories of mammals and their physical characteristics; then research to learn more about a mammal of their choice.

SUCCESS CRITERIA:

- identify three categories of mammals, describe their physical characteristics
- list different types of mammals
- choosing a mammal to research, describe how it is born and nurtured by its mother
- describe what it eats, how it moves, where it lives
- identify its predators
- describe any special changes the mammal experiences
- display all information using pictures and written descriptions

MATERIALS NEEDED:

- a copy of "Mammals" worksheet 1 and 2 for each student
- a copy of "In Terms of Mammals" worksheets 3 and 4 for each student
- a copy of "A Mammal Study" worksheets 5, 6, and 7 for each student
- access to the internet or local library
- pencils, pencil crayons, clipboards

PROCEDURE:

*This lesson can be done as one long lesson, or done in three or four shorter lessons.

1. Using worksheets 1 and 2, do a shared reading activity with the students. This will allow for reading practise and breaking down of the larger words. Along with content, discussion of some vocabulary would be beneficial for comprehension.

Some interesting vocabulary words to focus on are:

- egg-laying
- protection
- pouch
- appendages
- marsupial
- warm-blooded
- -female
- vertebrate
- placental
- nourish
- blindly
- underdeveloped

2. Divide students into pairs and give them worksheet 3. They will engage in a 'think-pair-share' activity to discuss the common characteristics of all mammals. A completed mind web would include these ideas:

- have fur or hair
- have a backbone
- feed their young milk
- have a skeleton inside their body
- give their young care and protection
- have 4 appendages
- are warm-blooded
- have lungs and need air to breathe

3. Continuing to work in pairs, give students worksheet 4 to complete.

4. Give students worksheets 5, 6 and 7. With access to the internet or by visiting a local library, students will find out more about a mammal of their choice.

DIFFERENTIATION:

lower learners may benefit by working as a small group with teacher support to complete worksheets 5, 6, and 7. Choosing one mammal, assign a section to each student in the group. Then they can share their findings with the small group in order to complete the worksheets.

For enrichment, faster learners could find out how long their chosen mammal lives in captivity (e.g., in a zoo), and in the wild. They could provide an explanation as to why there would be a difference in the life expectancy.

OTM2161   ISBN: 9781487710231
© On The Mark Press

Mammals

Did you know that there are about 5,400 different kinds of mammals in the world? These mammals can be divided into three main groups. There are egg-laying mammals, marsupials, and placental mammals.

Placental mammals give birth to live young. Mammals are the only animals that can feed their young milk.

Young chimpanzees stay close to their mothers until they are about eight or nine years old. Their mothers give them food, warmth, and protection.

When a mammal is born, it is not able to take care of itself. It needs its mother for food and protection. Some mammals need their mother's protection and care for only a few months, but others need care for many years.

Egg-laying mammals, such as the platypus and the echidna, lay eggs instead of giving birth to live young. When platypus babies hatch out of their eggs their mothers nourish their young with milk. They feed them for three to four months until they can swim on their own.

An echidna is like a spiny anteater. A female echidna lays an egg once a year, then ten days later it hatches. The baby echidna is called a puggle. The puggle will stay in its mother's warm pouch to feed on her milk for about another 50 days.

Marsupials are animals such as koalas, kangaroos, and opossums. These mammals give birth to live, but very underdeveloped young. The newborn baby mammal crawls blindly into the mother's pouch, where it will stay for the next few weeks or until it can go out on its own.

A baby kangaroo is called a joey. At birth a joey is only about the size of a grape. After it is born it crawls into its mother's pouch and stays there for about four months. It feeds on her milk. At around four months old, the joey will come out of its mother's pouch to feed on grass. At ten months, it is old enough to leave her pouch for good.

All mammals are **warm-blooded** animals. This means that they can control their own body temperature. They are **vertebrate** animals so they have a backbone and a skeleton on the inside of their body. They have lungs and need air to breathe. Mammals usually have four appendages (arms and legs), and they have large brains. All mammals are covered in fur or hair. Their hair or fur is used to keep them warm.

DID YOU KNOW?

Humans are mammals! We are placental mammals. Humans give birth to live young. We need our mothers for food and protection when we are young, until we learn to care for ourselves.

OTM2161   ISBN: 9781487710231
© On The Mark Press

In Terms of Mammals

Think **Pair** **Share**

With a partner, do some thinking and sharing of ideas about mammals.

Complete the web below by adding things that all mammals do or have in common with each other.

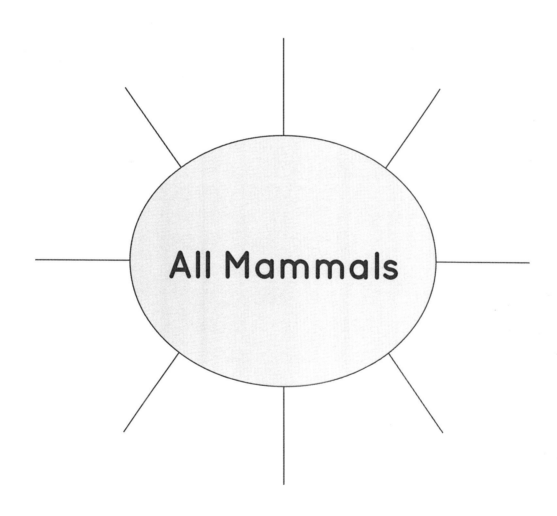

Name:

Keep Talking!

With your partner, brainstorm some examples of mammals. Use the box below to list or draw your ideas.

Circle one from the box that you would like to learn more about.

OTM2161   ISBN: 9781487710231
© On The Mark Press

Name:

A Mammal Study

Visit the library or access the internet to find out some interesting facts about a mammal that you would like to learn more about.

Choose It!

The mammal that I would like to learn more about is the

_____ .

Describe It!

Illustrate the mammal that you chose.

What type of mammal is it? (circle one)

egg-laying   marsupial   placental

OTM2161   ISBN: 9781487710231
© On The Mark Press

Study It!

How long does it take for it to hatch or be born?

_____

_____

_____

_____

How long does it stay with its mother?

_____

_____

_____

_____

What does it eat?

a) when it is a baby - _____

b) when it is a young mammal - _____

c) when it is an adult - _____

Circle all the ways that your mammal moves.

walk	run	hop
swim	fly	crawl

OTM2161   ISBN: 9781487710231
© On The Mark Press

Name:

A) Draw your mammal where it lives.

B) Draw and label some of its predators that live there too.

Tell about any special changes that happen to your mammal. For example:

- Does it migrate?

- Does it hibernate?

- Does it camouflage?

_____

_____

_____

OTM2161   ISBN: 9781487710231
© On The Mark Press

LIFE CYCLES

LEARNING INTENTION:

Students will learn about the life cycle stages of different animals.

SUCCESS CRITERIA:

- discuss the life cycle of a butterfly with a partner
- illustrate the changes that happen at each stage in a Monarch Butterfly's life
- order the stages in a frog's life and describe each stage in words
- discuss the life cycle of a frog with a partner
- read and discuss the life cycle of a mouse
- illustrate each stage of a mouse's life and describe each stage in words

MATERIALS NEEDED:

- a copy of "Life Cycle of a Monarch Butterfly" worksheet 1 and 2 for each student
- a copy of "Life Cycle of a Frog" worksheets 3, 4, and 5 for each student
- a copy of "Life Cycle of a Mouse" worksheets 6, 7, and 8 for each student
- access to the internet
- access to local library, sheets of art paper (*optional*)
- pencils, pencil crayons, glue, scissors

PROCEDURE:

*This lesson can be done as one long lesson, or done in three or four shorter lessons.

1. Do a read aloud about the life cycle of a butterfly. A suggested read aloud is <u>From Caterpillar to Butterfly</u> written by Deborah Heiligman. Afterward, discuss the stages of the life cycle of the butterfly with the students to ensure their understanding of the process, and the vocabulary. Students could engage in a 'turn and talk' activity. With a partner, they can talk about what piece of information from the story that they found most interesting/ what they still wonder about. Give students worksheets 1 and 2 to complete.

2. Allow students to watch a video about the life cycle of a frog. A suggested video is 'Wild Kratts Aqua Frog', which can be accessed at www.youtube.com. Afterward, discuss the stages of the life cycle of the frog with the students to ensure their understanding of the process, and the vocabulary. Students could engage in a 'turn and talk' activity. With a partner, they can talk about which piece of information from the video that they found most interesting/ what they still wonder about. Give students worksheets 3, 4 and 5 to complete.

3. Give students worksheet 6. Read the information through with the students to ensure their understanding of the content and vocabulary. Students can complete worksheets 7 and 8.

DIFFERENTIATION:

Slower learners may benefit by working as a small group with teacher support to re-read and discuss the information on worksheet 6. Highlighting important information, in a step by step process, will help these learners to recognize and record the information that they need in order to complete worksheets 7 and 8.

For enrichment, faster learners could chose an animal and investigate its life cycle. Dividing a sheet of paper into four quadrants will help these learners to organize their thinking and their output. Depiction of their animal's life cycle could be done using only pictures, or using pictures and written descriptions.

OTM2161   ISBN: 9781487710231
© On The Mark Press

Life Cycle of a Monarch Butterfly

Did you know that butterflies are not always butterflies for their whole life?

Read about the changes that happen at each stage of a Monarch Butterfly's life. Draw a picture of what is happening at each stage.

A female butterfly lays her eggs on the milkweed leaf. Each egg is the size of a pin head.

After about five days, a tiny caterpillar crawls out of an egg. The caterpillar is in the larva stage.

After two weeks of eating leaves, the caterpillar grows bigger. It has yellow, black, and white stripes.

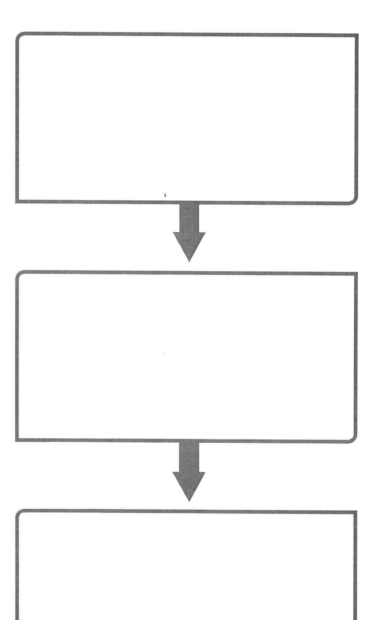

When it is fully grown the caterpillar attaches itself with silk to the bottom of a leaf. It hangs in the shape of a "J".

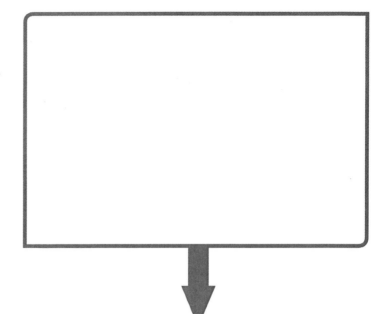

The caterpillar sheds its skin, leaving behind a green teardrop shape. It is now a pupa. The outside hardens and becomes a chrysalis.

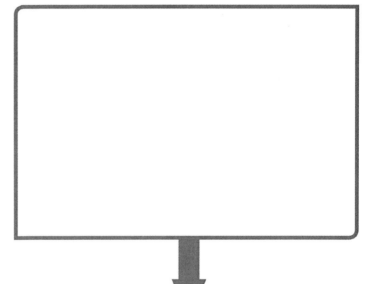

After ten days, the adult butterfly comes out. It has orange, black, and white stripes. The metamorphosis is complete!

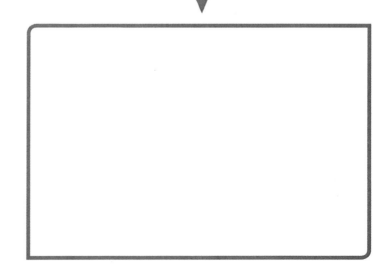

OTM2161   ISBN: 9781487710231
© On The Mark Press

Life Cycle of a Frog

Cut out the pictures below. On worksheets 4 and 5, order the pictures to show the life cycle of a frog. Beside each picture, write a sentence to explain what is happening at each stage of its life cycle.

_____

_____

_____

_____

_____

_____

_____

_____

_____

_____

_____

_____

_____

_____

_____

Worksheet 4

OTM2161   ISBN: 9781487710231
© On The Mark Press

Name:

Life Cycle of a Mouse

Mice are mammals and belong to the rodent family. They can be found almost all over the world. They are usually brown or gray in color. Mice are small animals. It is easy for them to get into all kinds of places!

Mice are fully adult when they are about two months old. They mate about four times a year. A female mouse will give birth to live babies. She will have about five to ten mice in a litter.

Did you know that newborn mice are blind, deaf, and have no hair? Their skin is pink in colour. At birth, mice are only about 2.5 cm long. The baby mice make little squeaking noises. The mother mouse feeds her babies milk for about the next week. The newborn mice cuddle up to keep warm.

After two days, the baby mice begin to get hair on their bodies. Their tiny whiskers begin to grow and their tails get longer. They can crawl and squeak, but they still cannot see or hear.

At about a week old, the hair of the baby mice grows thicker. Their ears pop away from their heads, and they are able to hear. But, their eyes are still closed!

By the time the baby mice are two weeks old, their eyes have opened and they can see. They are able to eat their own food and go exploring. They still cuddle up with their mother, and brothers and sisters to keep warm.

By four weeks old, the little mice are very active. They hop about, and they chase each other and play.

At eight weeks old, they are now adult mice. They can now start their own families.

OTM2161   ISBN: 9781487710231
© On The Mark Press

Name:

You have read about the life cycle of a mouse. Now, illustrate what is happening at each stage of its life. Beside each picture, write a sentence to explain what is happening.

Newborn

_____

_____

_____

_____

Day 2

_____

_____

_____

_____

_____

Day 7

_____

_____

_____

_____

OTM2161   ISBN: 9781487710231
© On The Mark Press

Name:

Continue illustrating what is happening at each stage of a mouse's life. Beside each picture, write a sentence to explain what is happening.

2 weeks old

_____

_____

_____

_____

4 weeks old

_____

_____

_____

_____

8 weeks old

_____

_____

_____

_____

OTM2161   ISBN: 9781487710231
© On The Mark Press

HUMAN DEVELOPMENT

LEARNING INTENTION:

Students will learn about the life cycle stages of a human, and about the healthy food that is needed for growth and development.

SUCCESS CRITERIA:

- recognize and record the changes that humans experience as they grow older
- recognize and record things that remain the same as humans grow older
- identify the four basic food groups
- plan a healthy daily menu for a child using appropriate serving sizes
- make observations and a conclusion about menu choices
- create a collage of healthy foods from the four food groups

MATERIALS NEEDED:

- a copy of "Life Cycle of a Human" worksheet 1, 2, and 3 for each student
- a copy of "Humans Need Food to Grow" worksheets 4, 5, and 6 for each student
- a copy of "What's On the Menu?" worksheets 7 and 8 for each student
- a copy of "A Count of Healthy Eating!" worksheet 9 for each student
- pencils, pencil crayons
- glue, scissors, art paper, old magazines

PROCEDURE:

*This lesson can be done as one long lesson, or done in four or five shorter lessons.

1. Give students worksheet 1. Read and discuss as a class how humans grow and change as they age.

2. Divide students into pairs. Give them worksheet 2. They will brainstorm with their partner ways in which a human being changes as it grows. An option is to come back together as a large group to share their responses. Responses could be recorded on chart paper.

3. Students will continue to work in pairs. Give them worksheet 3. They will brainstorm with their partner ways in which a human being stays the same as it grows. An option is to come back together as a large group to share their responses. Responses could be recorded on chart paper.

4. Follow-up with a group discussion about the fact that not all humans grow at the same rate, and that humans don't all have the same life span. Pose the question: 'What things could affect the length of a human life?'

5. Give students worksheets 4, 5, and 6. Read and discuss as a class what foods make up a healthy diet. A discussion about 'sometimes foods' would be appropriate too.

6. Give students worksheets 7 and 8. They will plan a daily menu for themselves that incorporates the appropriate daily servings from each of the food groups. Using worksheet 9, students will analyze their menu choices to see if they have met the suggested daily serving amounts. An option is to come back as a large group to discuss their findings. Pose the question, 'what changes in your diet do you think you need to make?'

7. Give each student a piece of art paper. Instruct them to fold it into four equal sections, then label each section as 'Fruits & Vegetables', 'Dairy', 'Grains', and 'Meat & Alternatives'. Using old magazines, students are to cut out and glue examples of healthy foods from each of the four food groups. This could be displayed on a bulletin board or around the school.

DIFFERENTIATION:

Slower learners may benefit by working with a strong peer to create a daily menu that incorporates healthy food choices, and appropriate serving sizes on worksheets 7 and 8.

For enrichment, faster learners could:

- be given worksheets 7 and 8 again, this time to plan a menu for their parents (remind students to look back at worksheet 5 to get the right serving sizes!)

- produce a piece of writing to attach to their healthy food collage, the writing should explain the importance of healthy eating and how it helps us grow and develop

OTM2161 ISBN: 9781487710231
© On The Mark Press

Life Cycle of a Human

People are human beings. We belong to the mammal group. Just like other mammals, we give birth to live young.

We feed our young babies milk and care for them until they are able to feed and care for themselves.

As humans, we too, have a life cycle. We are born. We are young children who grow into older children. We become adults, and sometimes we produce children of our own. Then, one day or another, we grow to be old.

Think Pair Share

With a partner, do some thinking and sharing of ideas about what changes in a human as they grow older. Record your ideas in the box.

OTM2161   ISBN: 9781487710231
© On The Mark Press

Think **Pair** **Share**

With a partner, do some thinking and sharing of ideas about what stays the same about a human as they grow older. Record your ideas in the box.

Humans Need Food to Grow

As humans, we need some basic things to live. We need water, air, shelter, and food. Eating good food and having a balanced diet is important to a healthy human life.

Eating foods from the four food groups will help us to grow. The four food groups are **grains**, **fruits and vegetables**, **meat and alternatives**, and **dairy**.

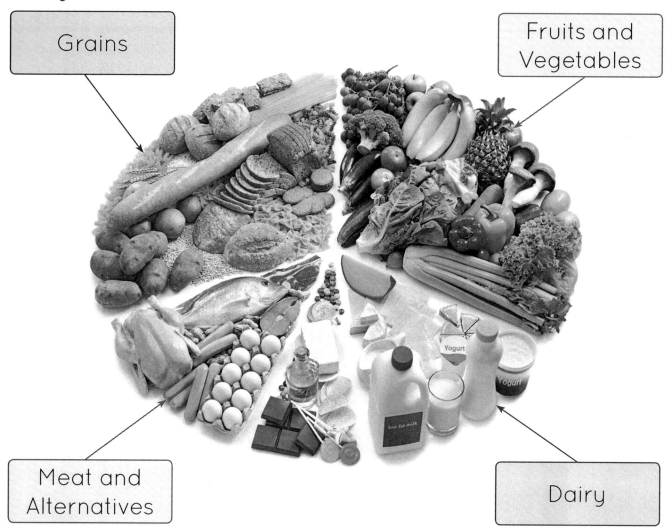

Grains

Fruits and Vegetables

Meat and Alternatives

Dairy

FAST FACT!

Foods like candy and chocolate are allowed, but only some of the time! That is why they are called 'sometimes foods'.

OTM2161   ISBN: 9781487710231
© On The Mark Press

Serve It Up!

Now that you know what kinds of foods make up a healthy diet, let's learn about how many servings you need from the food groups each day.

Food Groups	Children	Adults
Fruits and Vegetables	5 - 6 servings	7 – 8 servings
Grains	5 – 6 servings	7 – 8 servings
Dairy	2 – 3 servings	2 – 3 servings
Meat and Alternatives	1 – 2 servings	2 – 3 servings

DID YOU KNOW?

Eating a healthy balanced diet will help you to:

- Build strong bones and muscles
- Keep your brain alert
- Give you energy to work and play
- Keep your heart healthy
- Be a happier you!

OTM2161 ISBN: 9781487710231
© On The Mark Press

So what is a serving size?

Fruits and Vegetables

½ cup of vegetables

½ cup of fruit

½ cup of juice

Grains

1 slice of bread

½ cup cooked rice

¾ cup of cereal

½ cup cooked pasta

Dairy

1 cup of milk

¾ cup of yogurt

50 grams of cheese

Meat and Alternatives

75 grams of meat

2 eggs

¼ cup of nuts

¾ cup cooked beans

OTM2161   ISBN: 9781487710231
© On The Mark Press

What's On the Menu?

Plan a healthy menu for yourself. Be sure to fill it with a balanced diet by using foods from all food groups. Remember your serving sizes!

Breakfast

_____

_____

_____

_____

Snack time!

OTM2161   ISBN: 9781487710231
© On The Mark Press

Name:

Continue planning your menu.

Lunch

_____

_____

_____

_____

_____

Snack time!

Dinner

_____

_____

_____

_____

_____

_____

_____

OTM2161   ISBN: 9781487710231

Name:

A Count of Healthy Eating!

Look back at the menu you created. Did you meet your daily servings in all of the food groups? Let's check in! Record the number of servings that you ate from each of the food groups.

Food Groups	Number of My Servings
Fruits and Vegetables	_____ servings
Grains	_____ servings
Dairy	_____ servings
Meat and Alternatives	_____ servings

Did you get all of the daily servings from each food group? Explain.

_____

_____

OTM2161   ISBN: 9781487710231
© On The Mark Press

CO-EXISTING

LEARNING INTENTION:

Students will learn about how animals and humans co-exist on Earth and how they depend on each other for survival.

SUCCESS CRITERIA:

- identify animals that are useful to humans
- describe the ways that animals are useful to humans
- recognize ways that humans help and protect animals and their habitat
- recognize ways that humans and animals are harmful to each other
- describe how harmful incidents between humans and animals can be reduced
- create a food chain using pictures and written descriptions
- create a food web using pictures and written descriptions

MATERIALS NEEDED:

- a copy of "Animals and Us" worksheets 1, 2, and 3 for each student
- a copy of " Help Us!" worksheet 4 for each student
- a copy of "Humans Helping Animals" worksheet 5 for each student
- a copy of "Food Chains" worksheets 6 and 7 for each student
- a copy of "Make a Food Web" worksheets 8 and 9 for each student
- access to the internet or local library
- pencils, pencil crayons, markers, chart paper, scissors, glue

PROCEDURE:

*This lesson can be done as one long lesson, or done in five or six shorter lessons.

1. Divide students into pairs and give them worksheet 1. They will engage in a 'think-pair-share' activity to discuss what animals are useful to humans.

2. Give students worksheets 2 and 3. They will list the ways in which each animal is useful to humans.

3. Give students worksheet 4. Choosing an animal that they recorded on Worksheet 1, students will make a web to depict all the ways it is useful to humans. A follow-up option is to have students present their work to the large group, or in a small group. This will generate rich discussion about the different ways in which animals are useful to humans. Student work could be displayed on a bulletin board.

4. Divide students into pairs and give them worksheet 5. They will engage in a 'think-pair-share' activity to discuss ways in which humans are helpful and protective of animals and their habitat. Come back together as a large group. Pose these questions: What can cause a decline in animal population? What can happen if there is a decline in animal population?

5. Engage students in a group discussion about the ways in which animals can be harmful to humans, and the ways in which humans can be harmful to animals. Answers can be recorded on chart paper. Pose this question: How can these harmful incidents be minimized?

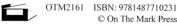
OTM2161 ISBN: 9781487710231
© On The Mark Press

6. Give students worksheet 6. Read the information and discuss to ensure students' understanding of the material. Give students worksheet 7 to complete. Students may need access to the internet or a local library to get information to complete their food chain. A follow-up option is to have students present their food chain within a small group. This will enhance their knowledge of different food chains that exist on our Earth, and reinforce the understanding that every living thing has a purpose.

*As a further activity to enhance the learning about food chains, show students The Magic School Bus episode called "Gets Eaten". Accessing the internet, go to www.youtube.com for the full episode. A follow-up option is to have students illustrate the food chain in this episode.

7. Broaden students' knowledge by discussing food webs. Explain that all plants and animals are interconnected. Many connections exist within a small group of plants and animals. Using worksheets 8 and 9, students will create a food web. Students may need access to the internet or a local library to get information to complete their food chain. Food chains could be displayed on a bulletin board. A follow up option is to have students explain their food web in words, then attach it to their food web display.

DIFFERENTIATION:

Slower learners may benefit by working with a partner to make a food web.

For enrichment, faster learners could add plants and animals of their choosing to the food web by drawing them in the empty spaces on the page, and then draw their connections to the other plants and animals in the existing food web. They could also choose to create another food web of their own.

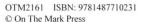

Animals and Us

There are a lot of different species on our planet Earth. Humans and animals have to share the Earth. In fact, all the different species on Earth need each other to live and grow. This makes us useful to each other.

Think Pair Share

With a partner, do some thinking and sharing of ideas about what animals are useful to humans. Record your ideas in the box.

OTM2161 ISBN: 9781487710231
© On The Mark Press

Name:

Many animals help humans by giving us food, material for clothing, fertilizer for our gardens, and even being friendly pets that make us happy.

In the chart, explain how each animal is useful to humans. Some animals may have more than one use!

Animal	How it is useful to humans
sheep	
cow	
fish	
chicken	

OTM2161   ISBN: 9781487710231
© On The Mark Press

Name:

In the chart, explain how each animal is useful to humans. Remember that some animals may have more than one use!

Animal	How it is useful to humans
goat	
bee	
worm	
horse	
dog	

OTM2161   ISBN: 9781487710231
© On The Mark Press

Name:

Help Us!

Choose one of the animals that you recorded on Worksheet 1, write it in the box title. Now, draw a picture of it in the centre of the web. Complete the web to show all the ways it is useful to humans.

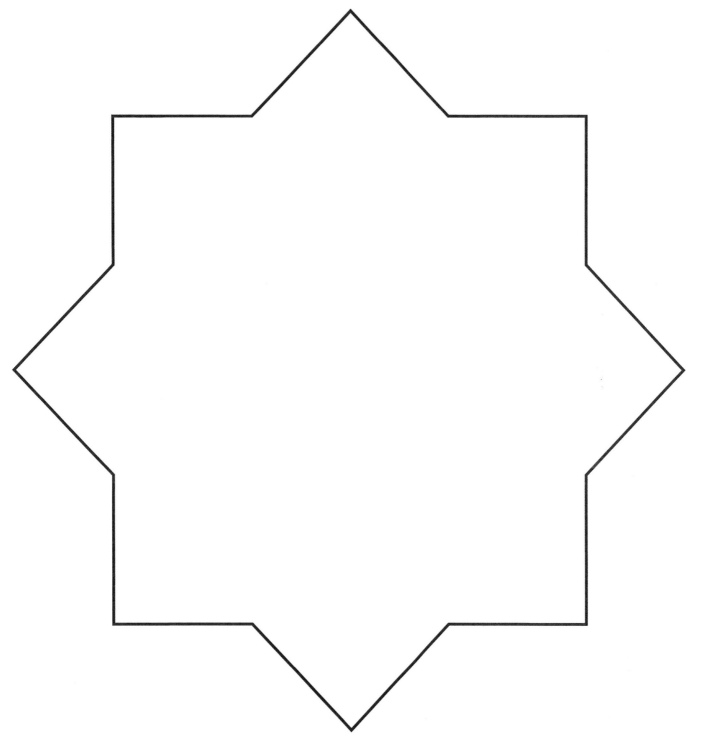

Name:

Humans Helping Animals

Think **Pair** **Share**

Humans can be helpful to animals too. With a partner, do some thinking and sharing of ideas about the ways that humans help and protect animals and their habitat. Record your ideas in the box.

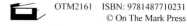

OTM2161   ISBN: 9781487710231
© On The Mark Press

Food Chains

You have learned that animals can be useful to humans. Did you know that animals are also useful to each other?

Animals get energy from plants and other animals because they provide food for each other. They are all part of a food chain. Let's learn more about this!

So, it all begins with the Sun. You see, the Sun's heat energy makes plants grow. Plants are food for some animals. Some animals are food for each other.

This is a food chain. Grass is food for a grasshopper. A grasshopper is food for a mouse. A mouse is food for a snake. A snake is food for a hawk.

OTM2161   ISBN: 9781487710231
© On The Mark Press

Name:

It is your turn to create a food chain. In the boxes, draw the living things that are part of a food chain. Start with the Sun!

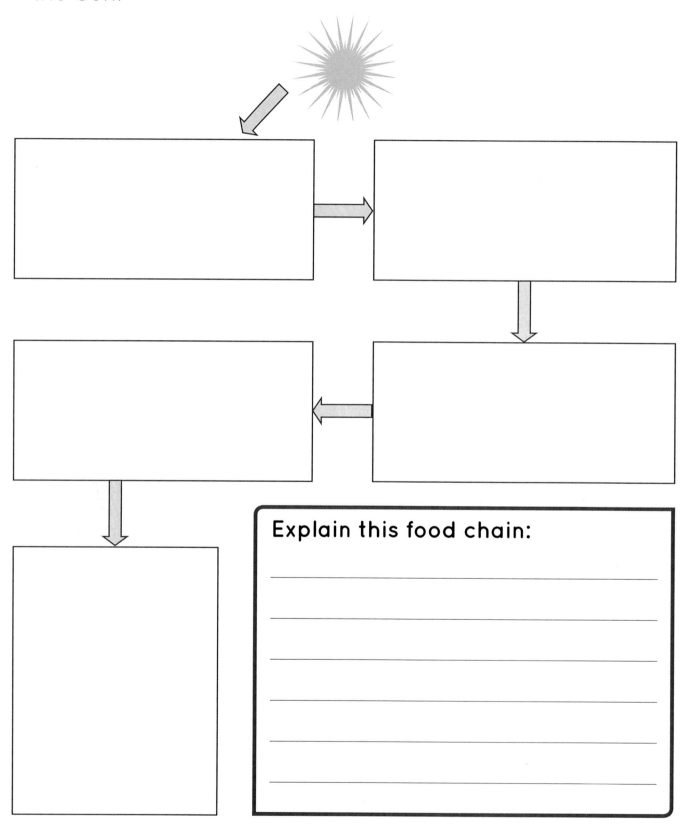

Explain this food chain:

_____

_____

_____

_____

_____

_____

OTM2161   ISBN: 9781487710231
© On The Mark Press

Name:

Make a Food Web

Food chains are all connected. A food web is a bunch of food chains. Let's create a food web!

Materials Needed:

- scissors

- glue

What To Do:

1. Cut out the pictures on worksheet 8.

2. On worksheet 9, glue the pictures to make a food web.

3. Draw lines between plants or animals that eat each other.

Why is it called a food web and not a food chain?

_____

_____

_____

_____

OTM2161　ISBN: 9781487710231
© On The Mark Press

AN ABORIGINAL CONNECTION

LEARNING INTENTION:

Students will learn about the importance of certain animals to the lives of Aboriginal people.

SUCCESS CRITERIA:

- discuss with a partner the usefulness of some animals to Aboriginal people
- depict the uses of animals to Aboriginal people of long ago
- conduct an interview with an Aboriginal Elder to learn about the importance of some animals in their culture today
- use drawings and written descriptions to indicate responses

MATERIALS NEEDED:

- a copy of "Long, Long Ago" worksheet 1 for each student
- a copy of "Animals in Aboriginal Lives" worksheets 2 and 3 for each student
- access to the internet or local library
- pencils, pencil crayons, clipboards
- assorted paint colors, paint brushes, white art paper or poster paper *(optional materials)*

PROCEDURE:

*This lesson can be done as one long lesson, or be divided into three shorter lessons.

1. Do a read aloud about how Aboriginal people use and respect animals. A suggested read aloud is Little Water and the Gift of Animals written by C.J. Taylor. Afterward, discuss ways that animals have been useful to Aboriginal people and their lifestyle. Also, a discussion about the thoughts and feelings that Aboriginal people have toward certain animals would be beneficial. Students could engage in a 'turn and talk' activity. With a partner, they can talk about what piece of information from the story that they found most interesting/ what they still wonder about. Asking some pairs to share their thoughts with the whole group would promote rich discussion.

2. Provide students with some resources to research some ways that animals were used by Aboriginal people long ago (and perhaps still are today). Students can use this information to complete worksheet 1.

3. **Invite an Aboriginal Elder from a local community** to come in to speak to the students about the animals that are important to Aboriginal people. Give each student a clipboard, pencil, and worksheets 2 and 3 to complete as they conduct the interview.

DIFFERENTIATION:

Slower learners may benefit by working with a partner to research and complete worksheet 1.

For enrichment, faster learners could:

- choose one animal of importance to Aboriginal culture, draw it, then label all the parts of the animal that are used
- draw and paint a picture of how one of the animals from the story Little Water and the Gift of Animals helps the people of the village
- prepare three additional questions to ask the Elder

OTM2161   ISBN: 9781487710231
© On The Mark Press

Name:

Long, Long Ago

Construct a mind map to show how Aboriginal people who lived long ago used animals in their daily lives. Use pictures and words in your map.

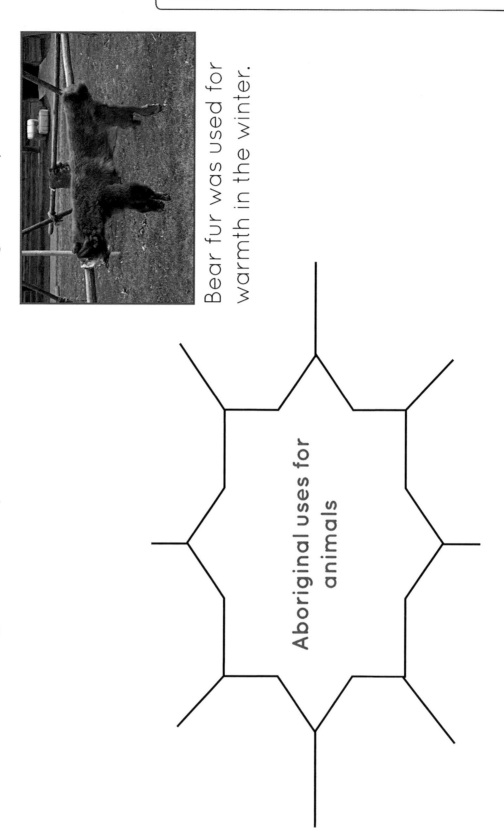

Bear fur was used for warmth in the winter.

Aboriginal uses for animals

94

OTM2161   ISBN: 9781487710231
© On The Mark Press

Animals in Aboriginal Lives

Interview an Aboriginal Elder. Use the questions below as a guide when asking the Elder about the importance of animals to his / her people today.

Q. Name an animal that is important to Aboriginal people.

A. _____

_____

_____

Q. What are some of the uses of this animal to Aboriginal people?

A. _____

_____

_____

Q. Name two other animals that are important to Aboriginal people in your area. What do you use them for?

A. (draw and label the Elder's answers)

Use the questions below as a guide when asking the Elder about the importance of animals to his / her people today.

Q. What animals are important in Aboriginal celebrations?

A. (draw and label the Elder's answers)

Q. Explain the importance of one of these animals to an Aboriginal Celebration.

A. _____

_____

_____

OTM2161   ISBN: 9781487710231
© On The Mark Press